Living Things and Their Environments

by Laura McDonald

Table of Contents

Introduction
Pack Your Bags for a Space Trip! 4
You're taking a trip into space!
What will you need to take along?

Chapter 1
Environments and Ecosystems 6
How does the environment affect living things?
Cartoonist's Notebook 14

Chapter 2
Food and Energy in the Environment . . . 16
What paths do food and energy follow in an ecosystem?

Chapter 3
Cycles in Ecosystems 26
How do ecosystems recycle important chemicals?

Chapter 4
Changes in Ecosystems 34
How do ecosystems change over time?

Conclusion
Biosphere 2 42

How to Write a Press Release 44
Glossary 46
Index 48

How do living and nonliving things interact and maintain balance in the environment?

Introduction

Pack Your Bags for a Space Trip!

Congratulations! You've just won a trip to space! You'll be traveling in the space shuttle to the International Space Station. It's time to pack your bags.

First, pack the things you would need on any trip. Pack your clothes and your toothbrush. Pack magazines, too. You will also need to pack food. Outer space does not have any food. This is because no plants or animals live in space.

Why are there no living things in space? One reason is that it is very cold! Outer space also has no air. No air means no oxygen to breathe. So you have to also pack all the oxygen you will need on your trip.

Your body will convert, or change, that oxygen into carbon dioxide. Too much of this gas is poisonous. On Earth, plants take in the extra carbon dioxide. In space, you will need air filters to do that job.

You will have to pack enough water to last the whole trip. This is because outer space does not have liquid water. You will also have to think about what to do with your body wastes: either store them or release them into space.

Wow! This is a lot of packing—nearly 1,000 kilograms (2,200 pounds) of gear. We often take it for granted, but packing for a trip on Earth is easy.

That is because Earth gives each living thing what it needs to survive, or stay alive. Read this book to learn how different living things interact with their environments.

Chapter 1

Environments and Ecosystems

HOW DOES THE ENVIRONMENT AFFECT LIVING THINGS?

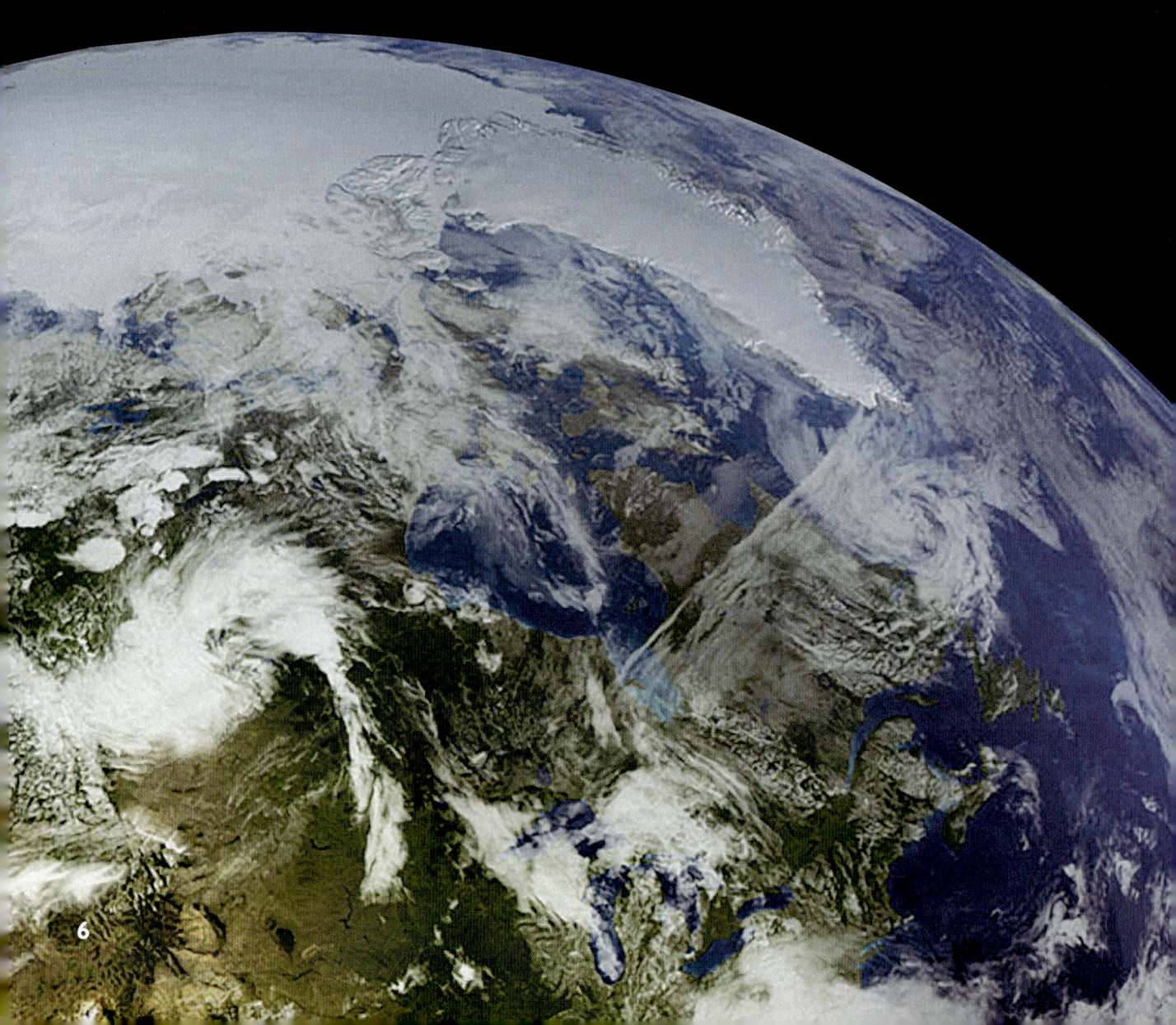

ESSENTIAL VOCABULARY

abiotic factor	page 11
biosphere	page 7
biotic factor	page 11
community	page 8
ecology	page 8
ecosystem	page 8
habitat	page 10
niche	page 12
population	page 10

Imagine you are at the International Space Station. You are looking down at Earth. Earth is a ball of rock heated and lit by a star. That star is the sun. A thin layer of air surrounds Earth. This air contains the gases. Clouds of water droplets move slowly around the planet. These clouds float over the colored patches on Earth's surface. The brown and green patches are land. The blue patches are water.

These patches are home to the thin layer of living things that covers Earth. The living things come in all shapes and sizes. Huge trees grow on Earth. Giant whales swim in the ocean. Millions of species, or types, of insects also live here.

THE SPHERES OF EARTH

You have just taken a tour of Earth's four major spheres. The air and gases make up Earth's atmosphere. The land and rock in Earth's crust make up the lithosphere. The water on Earth makes up the hydrosphere. And all the living things on Earth make up the **biosphere** (BY-uh-sfeer).

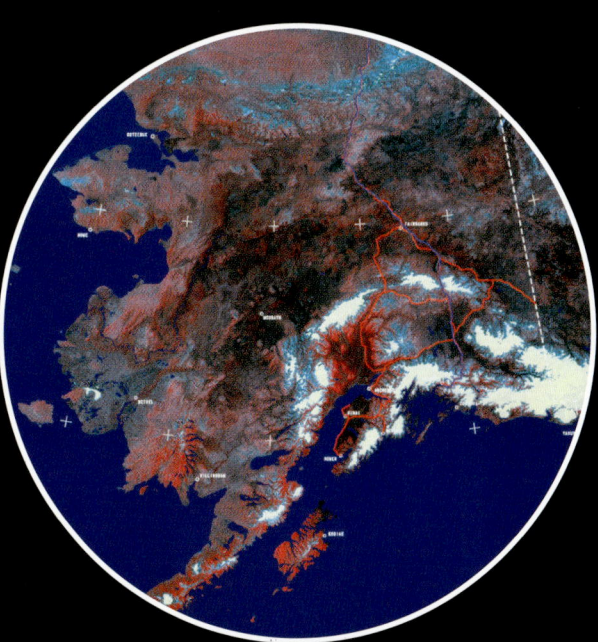

▲ From space, Alaska looks like a patchwork of colors.

Chapter 1

Earth's spheres are all connected. Most of Earth's creatures live in the water. Others need water to carry nutrients throughout their bodies and remove waste.

The atmosphere contains many gases. These gases protect Earth from the sun's harmful rays. These gases also help keep Earth warm. These gases allow living things to survive. We need the gases oxygen, carbon dioxide, and nitrogen to live.

The minerals in Earth's crust help living things grow and reproduce. Plants take in minerals from Earth's crust. Then animals eat the plants. Living things and their nonliving environments are all connected.

ENVIRONMENT AND ECOSYSTEM

We can use a weasel as an example. Think of a weasel hunting in a rocky meadow. All the living and nonliving things around the weasel are its environment. Air and sunlight are part of the environment. Rocks are part of its environment, too. Plants in the meadow and fleas in the weasel's fur are part of the environment. The mouse the weasel will eat for dinner is also part of the environment. **Ecology** (ih-KAH-luh-jee) is the study of the relationships among living things and their nonliving environments.

WHAT IS AN ECOSYSTEM?

Ecologists call the places they study **ecosystems** (EE-koh-sis-tumz). An ecosystem is all the living and nonliving things in one place. All of these living and nonliving things interact with one another. Ecosystems can be large or small. An ecosystem could be as small as a tree stump. An ecosystem can also be Earth itself.

COMMUNITIES OF LIVING THINGS

All the living things in an ecosystem form a **community** (kuh-MYOO-nih-tee). A pond can be an ecosystem. The community would be all the animals and plants that live in and near the pond. The community includes worms and molds. The community also includes bacteria.

Science & Math

Often scientists cannot make an exact count or measurement of what they study. Instead, they estimate, or make a best guess, based on the available evidence. Right now, scientists estimate that they have described 10 to 15 percent of the species on Earth. With 1.75 million types described so far, there could be at least 17.5 million species on Earth!

Environments and Ecosystems

Atmosphere (air)

Lithosphere (earth)

Biosphere

Hydrosphere (water)

The **Root** of the Meaning

Environment comes from the French word *environs*, meaning "surroundings."

▲ An ecologist may study a forest ecosystem or the ecosystem of a single dead tree.

Chapter 1

Each community member's actions affect the others. For example, fish eat insects and insects eat plants. Bacteria break down the dead bodies of plants and animals. The nutrients from the bodies are then available for use by plants.

A community contains many populations. A **population** (pah-pyuh-LAY-shun) is all the members of one species in a certain place. Three populations in a pond might be box turtles, snapping turtles, and cattail plants.

HABITATS

Each living thing has a **habitat** (HA-bih-tat), or place where it lives. Cattail plants live in ponds. Horned lizards live in the desert. Norway rats live in buildings and boats. Habitats provide food, water, and homes for living things.

Some habitats are in danger. These habitats are becoming less common. For example, the snowy polar habitat is shrinking as the world warms. The living things in this habitat may die out.

Have you ever visited a zoo? Have you seen the habitats people make for each animal? In a zoo, the polar bears swim in a cold pool with chunks of ice in the water. The zoo has rocks for them to climb. Zookeepers know that they must copy an animal's natural habitat to keep it alive.

▲ Many biotic and abiotic factors affect this weasel.

▲ All the members of one species in a community make up a population. These coral polyps are a population.

CHECKPOINT ✓

VISUALIZE IT

Draw and label four circles to represent the four levels of ecology: biosphere, ecosystem, community, and population. Inside the biosphere circle, make smaller circles showing several ecosystems. Inside the ecosystem circle, label smaller circles showing the biotic community and abiotic factors. Add small circles in the community circle for populations, and in the population circle for individuals.

Environments and Ecosystems

BIOTIC AND ABIOTIC FACTORS

A living thing is affected by both living and nonliving factors in its environment. **Abiotic factors** (ay-by-AH-tik FAK-terz) are nonliving factors. These factors include weather, air, and soil. Weather, for example, determines if mice are running around above the ground where the weasel can hunt them.

Biotic factors (by-AH-tik FAK-terz) are living factors. Living factors that affect the weasel might include hawks, mice, grass, and fleas. Hawks hunt weasels, forcing them to stay in the tall grass to survive. The weasel in turn eats mice that live in the grass. Meanwhile, fleas in the tall grass will drink the weasel's blood.

Biotic and abiotic factors differ from one ecosystem to another. Toxic metals from a mine may pollute one valley and poison the local plants. In that valley, there will be few seeds for mice to eat and little food for weasels. All of the factors in an ecosystem combine to determine the types and numbers of living things found there.

◀ The habitat for polar bears is the icy north.

Chapter 1

NICHES

A habitat has many types of living things. Each species has a strategy for survival. This strategy is called a **niche** (NICH). The niche of a weasel is to hunt small animals both on and in the ground. The shape and behavior of living things help them succeed. A weasel has a long, skinny body. This allows it to run through narrow tunnels in the ground. A weasel kills all the prey it can. Then it stores the extra food in its den for another day.

All the living things in a habitat share the same resources. But each species has a different niche. For example, several types of grazing animals can live together in a grassland. Zebras eat the stems of long grasses. Wildebeests bite off the stems on the sides of grass plants. Thompson's gazelles eat low plants that grow near the ground. All three can share one patch of grass because of their different niches.

Thompson's gazelle

zebra

wildebeest

> **CHECKPOINT** ✓
> **THINK ABOUT IT**
> How is a living thing's niche like a person's job? How is it different?

◀ Clown fish hunt for tiny animals.

▲ Anemone attach to coral and eat fish and shellfish that they sting to death.

Environments and Ecosystems

Summing Up

- Earth contains four interacting systems: the atmosphere, hydrosphere, lithosphere, and biosphere.

- All the living things on Earth make up the biosphere. Living things are influenced by biotic (living) and abiotic (nonliving) factors in their environments.

- Ecology is the study of the interactions among living things and their nonliving environments.

- Ecologists divide the biosphere into ecosystems. Ecosystems contain nonliving factors and living communities.

- A group of one kind of living thing in a given location is called a population.

- Each living species has a habitat and a specific niche.

Putting It All Together

Choose one of the activities below. Work independently, in pairs, or in a small group. Share your responses with the class. Listen to other groups present their responses.

1. How could scientists create a habitat for plants to grow on the International Space Station? Make a poster that shows how the needs of the plants could be met. Use what you have learned in the chapter, and do additional research if necessary.

2. People put many types of plants in their gardens. Reread the chapter and explain why some plants stay in their garden beds while others spread into neighboring yards and fields.

3. Draw a picture of the scene you see outside your window. Label the members of the natural community and the biotic and abiotic factors that affect them.

What is your niche?

Are you...
- an herbivore?
- a carnivore?
- an omnivore?

Humans are top consumers. Why do you think this is? What do you think it would be like to not be at the top?

Chapter 2

Food and Energy in the Environment

WHAT PATHS DO FOOD AND ENERGY FOLLOW IN AN ECOSYSTEM?

Living things are always growing, moving, and reproducing. Each of these actions requires energy. How do living things get energy?

ENERGY DRIVES ECOSYSTEMS

Earth's energy comes from the sun. Many people think that the dinosaurs died from lack of sun. People think that an asteroid hit Earth 65 million years ago. The impact of this rock from space caused a dust cloud. The dust cloud blocked some of the sun's light. This started a chain of events that killed off 75 percent of the living things on Earth.

Why is sunlight so important? Plants and algae use the sun's energy to make food. This process is called photosynthesis. The food powers their lives and builds their bodies. Living things that can make their own food are called **autotrophs** (AU-tuh-trofes). Autotrophs are also called **producers** (pruh-DOO-serz). This is because they produce food for other living things.

ESSENTIAL VOCABULARY

autotroph	page 16
carnivore	page 18
consumer	page 18
decomposer	page 18
food web	page 21
herbivore	page 18
omnivore	page 18
producer	page 16

Chapter 2

▲ Decomposers eat the dead bodies of former living things.

EVERYTHING GETS EATEN

Living things that cannot produce their own food are **consumers** (kun-SOO-merz). Consumers eat producers to get energy. They may also eat other consumers. All animals are consumers.

We can divide consumers into three main types. **Herbivores** (ER-bih-vorz) are animals that eat plants. **Carnivores** (KAR-nih-vorz) are animals that eat other animals. **Omnivores** (AHM-nih-vorz) eat both plants and animals.

Every living thing makes waste. In time, every living thing also dies. The waste and dead bodies of living things are food for **decomposers** (dee-kum-POH-zerz). Some mushrooms eat dead wood. Dung beetles eat cow droppings. Tiny mites eat dead skin cells. Why eat waste? Eating waste is a good source of energy. Waste is also easy to find. The waste of decomposers then helps plants grow.

Science to Science: Earth Science + Biology

Some ecosystems in the deep ocean are powered by heat and chemicals that well up from the ocean floor. Unusual animals such as tube worms and eyeless shrimp graze on bacteria that live in these dark ecosystems.

Food and Energy in the Environment

HANDS-ON SCIENCE: Observe a Decomposer

Decomposers such as molds eat dead plants and animals in every ecosystem. The spores of mold are in the air around us. What types of materials will molds decompose?

TIME REQUIRED
20 minutes, plus 10 minutes per day for 5 days

MATERIALS NEEDED
- five reclosable snap-and-seal bags
- three food samples, such as bread, fruit, and dairy product
- piece of green plant
- small piece of cardboard
- packing or duct tape

SAFETY CONSIDERATIONS
Do not open bags after you seal them. Do not eat the food samples.

PROCEDURE
1. Place each food sample in a plastic bag. Put the plant sample in the fourth bag and the cardboard in the fifth bag. Moisten each sample with a few drops of water. Label each bag with what is inside.
2. Snap the bags shut. Use tape to seal the entire top of each bag.
3. Describe the contents of each bag in a data table like the one shown below. Put the bags in a warm, dark place.
4. On each of the next five days, describe the contents of each bag. At the end of the experiment, put the bags directly in the trash without opening them.

ANALYSIS
1. Compare and contrast the appearance of each sample at the beginning and end of the experiment. Write or draw your observations.
2. Based on your results, what kinds of material do molds decompose?

	Food Sample 1	Food Sample 2	Food Sample 3	Plant	Cardboard
Start					
Day 1					
Day 2					
Day 3					
Day 4					
Day 5					

Chapter 2

They Made a Difference

Biruté Galdikas

(1946–)

▲ Biruté Galdikas with an orphaned orangutan.

Can you imagine devoting your life to the study of one type of animal in a faraway place? That's what Biruté Galdikas (bih-ROO-tay GAHL-dih-kus) has done. She moved from the United States to Indonesia in 1971 to study orangutans. When Galdikas arrived, people knew very little about the lives of wild orangutans.

Galdikas followed orangutans through the jungle and learned that they eat over 400 foods, including fruit, insects, honey, and bark. Baby orangutans stay with their mothers for about nine years, learning how to find and eat these foods. Other than mothers with babies, orangutans lead a solitary life.

The habitat for orangutans is the wild jungle of Indonesia and Malaysia. Unfortunately, people are cutting down that jungle to get lumber and land for palm plantations. There are about 50,000 to 60,000 orangutans left in the world. Galdikas worries that they will lose their habitat and die out in the next twenty years. She now spends much of her time working to preserve the jungle for all the species that make it their home.

Careers in Science: Naturalist

Naturalists help people learn about ecosystems. Many naturalists work in nature centers. Naturalists love the outdoors and enjoy teaching children and adults about nature. Naturalists learn by spending time outdoors, taking classes, and working with other naturalists. Ask at a local park how you can become a junior naturalist!

Food and Energy in the Environment

FOOD WEB

The energy in your body comes from the food you eat. Energy moves through ecosystems. Energy from the sun moves through plants. Next, this energy moves through animals that eat plants. Then this energy moves through animals that eat other animals. This forms a food chain. A simple food chain may have a palm tree, a beetle that eats palm fruits, and an orangutan (uh-RANG-uh-tan) that eats beetles.

Ecosystems usually have many food chains. A **food web** shows how food chains interact. A food web is a diagram. A food web shows how energy moves through an ecosystem. Look at the food web on this page. The arrows show where each member gets its energy. The food web shows the relationship between producers and consumers. Predators are consumers that eat other animals. Prey are the animals that are eaten.

Most communities have thousands of species. These species all eat one another. A food web simply shows a few of the ways energy moves through the community.

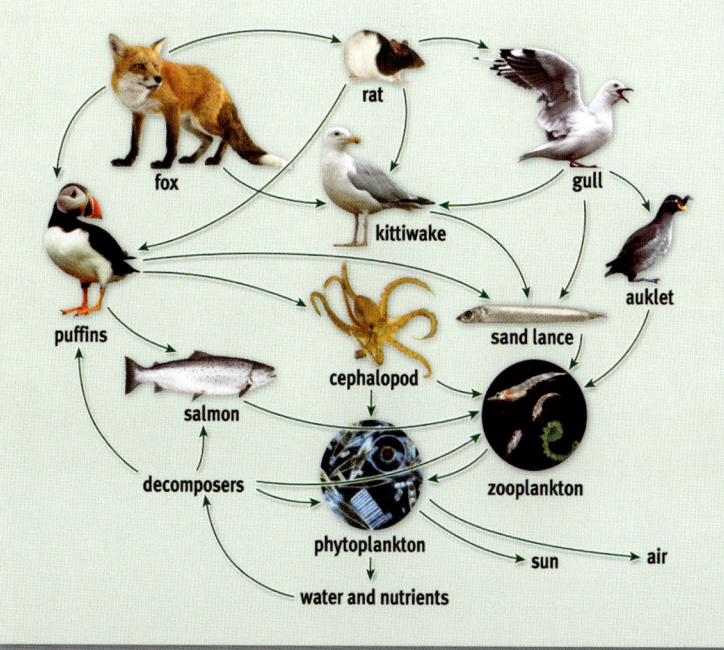

CHECKPOINT ✓

MAKE CONNECTIONS

Many cats live outside and hunt for their own food. Where do you think a cat would fit in the coastal marine food web?

21

Chapter 2

LIMITS ON LIFE

When you want something to eat, what do you do? You can look in your kitchen. You can buy food at a store. You can even order a pizza. Consumers in an ecosystem have fewer options. Consumers must eat the producers and other consumers in their community.

Producers are the base of every food web. The food available to the consumers in an ecosystem is the biomass, or total mass, of the producers. Some ecosystems have more producers than others. Warm, wet places like rain forests are covered in thick layers of producers. Many consumers can live on the biomass of plants there. Cold or very dry places have few producers. As a result, those places have fewer consumers and simpler food webs.

Consumers cannot digest all of a plant's biomass. The energy they can get is used for different things. Consumers use part of the plant's energy to keep warm. The rest of the energy helps the consumer move, grow, and reproduce.

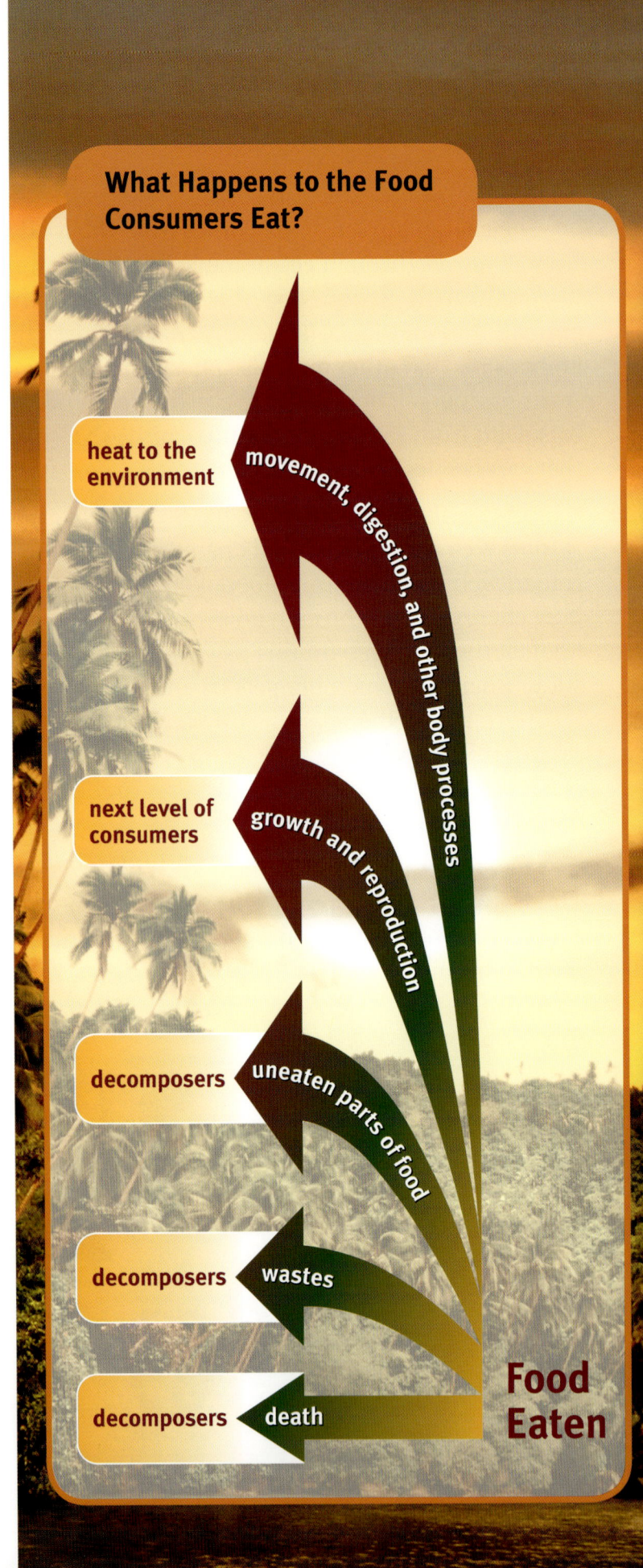

What Happens to the Food Consumers Eat?

- heat to the environment — movement, digestion, and other body processes
- next level of consumers — growth and reproduction
- decomposers — uneaten parts of food
- decomposers — wastes
- decomposers — death

Food Eaten

22

Chapter 2

When an herbivore eats a plant, less than ten percent becomes new herbivore biomass. In turn, carnivores convert less than ten percent of the biomass they eat. This means the food available to each level of a food chain shrinks from producer to herbivore to carnivore. That is why there are more plants and insects than tigers and killer whales. Look at the diagram on this page. You can see how biomass shrinks as it moves through a food chain.

Science to Science: Physics + Biology

The energy lost between each step of a food chain follows the Second Law of Thermodynamics. This law states that in any energy conversion, some energy is lost as heat. Batteries, lightbulbs, and car engines are subject to the same limitations as life.

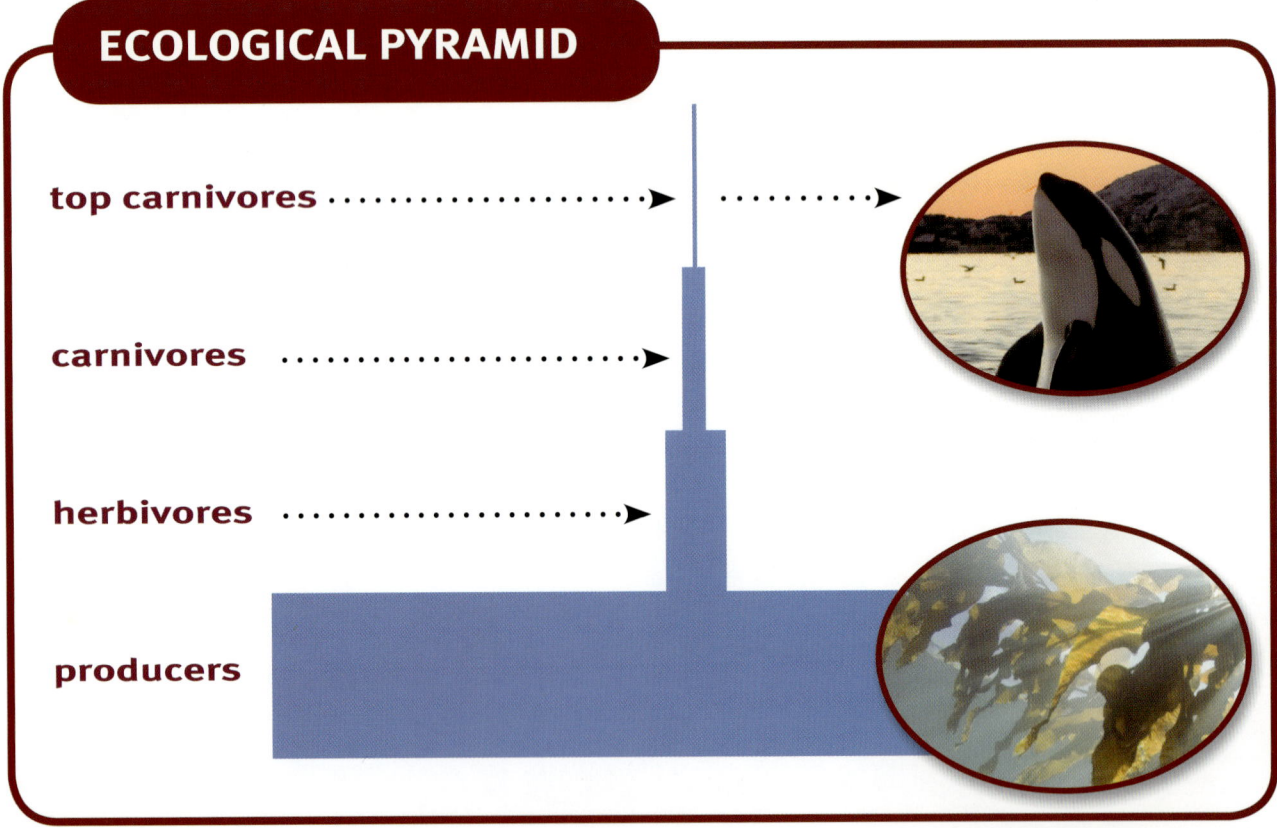

ECOLOGICAL PYRAMID

- top carnivores
- carnivores
- herbivores
- producers

▲ Herbivores have much more food available to them than top carnivores.

Food and Energy in the Environment

Summing Up

- Producers make their own food using sunlight and basic chemicals in the environment.

- Consumers get their energy by eating producers or other consumers.

- Herbivores, carnivores, omnivores, and decomposers all live on the energy harnessed by producers.

- Food chains and food webs show how energy moves from producers up through the different levels of consumers.

- A lot of producer biomass is needed to sustain primary and secondary consumers because energy is decreased in each transfer. This relationship may be shown using a pyramid.

Putting It All Together

Choose one of the activities below. Work independently, in pairs, or in a small group. Share your responses with the class. Listen to other groups present their responses.

1. People often leave decomposers out of food chains and food webs. Write a science fiction story about what would happen if there really were no decomposers on Earth.

2. Some people worry that there will not be enough food to support the growing human population. Based on the information on pages 22–24, would meats or plant foods feed the most people? Explain your answer. Make a poster illustrating your conclusions.

3. Learn more about food webs in the Arctic, desert, or rain forest. Make a poster showing one of these food webs.

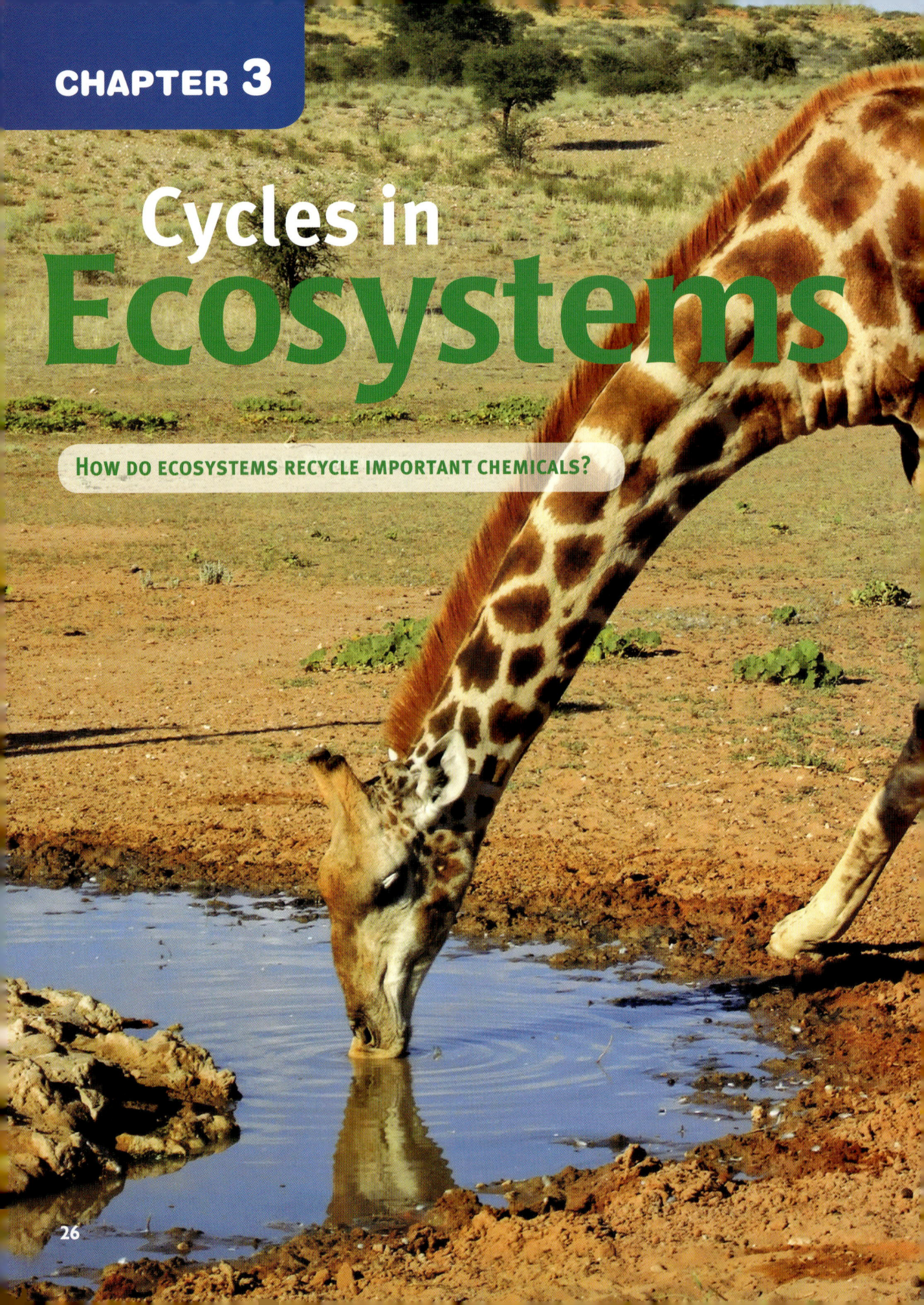

CHAPTER 3

Cycles in Ecosystems

How do ecosystems recycle important chemicals?

▲ The water cycle ensures that living things do not run out of water.

The environment gives living things what they need. The air, water, and soil provide the gases and nutrients living things use to live. Why is it that Earth's supply of these things does not run out? How is it that we can rely upon a steady supply of oxygen here, but astronauts in outer space cannot?

The answer lies in the cycles of nature. Some processes use the important substances, while other processes make them. Some living things put nitrogen into the soil. Others send it back into the air. Water dries up from the land and then falls back down as rain and snow. Consumers and producers take oxygen out of the air. But producers also put more oxygen back into the air. These cycles allow life on Earth to continue.

Chapter 3

THE NITROGEN CYCLE

Living things need nitrogen to grow. Nitrogen travels from the air to soil and then through the food web.

Look at the diagram below. This diagram shows the nitrogen cycle. Bacteria in the soil convert nitrogen to compounds that plants can use (A). Next, plants use the nitrogen to build proteins and grow (B). Animals eat some plants (C). Animals recycle the proteins in the plants into animal proteins.

Waste and dead matter fall to the ground (D). Here decomposers change them into simpler compounds. These compounds are called fixed nitrogen compounds (E). The converted, or fixed, nitrogen is now ready for new plants to use. Other bacteria also return nitrogen to the atmosphere (F). The end result is that fixed nitrogen is again ready for living things to use. This is how the level of nitrogen in the atmosphere stays constant.

▲ The lumps on these roots contain nitrogen-fixing bacteria.

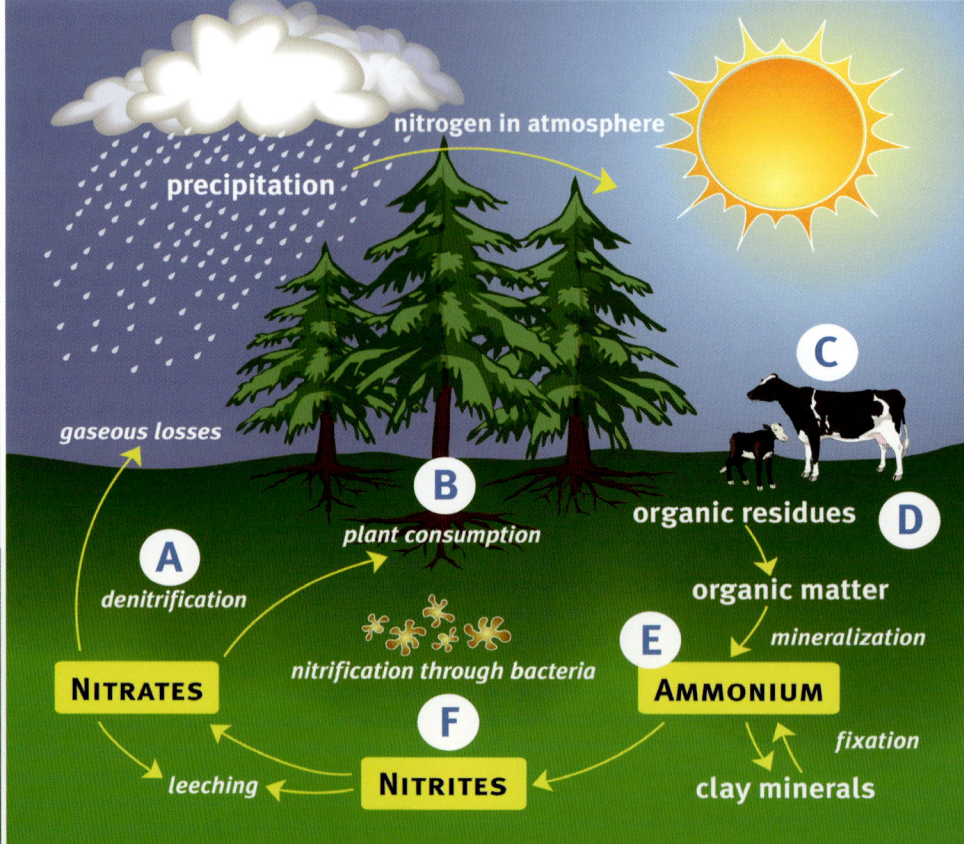

28

Cycles in Ecosystems

THE WATER CYCLE

Life cannot exist without water. Many chemical reactions need water. Water also carries substances through living things. Look at the diagram below. The water cycle shows how water moves through the air, land, and living organisms.

Water comes out of the air in the form of precipitation (A). This rain or snow soaks into the ground or runs off into ponds, streams, lakes, and oceans (B). Living things take up water from the land and from bodies of water (C). Some of the water evaporates into the air. This means the water becomes a gas called water vapor (D). Water also evaporates from plants during transpiration (E).

Moist air rises. The water vapor condenses into clouds of water droplets (F). Then the droplets fall back to Earth as rain or snow. Every drop of water on Earth has passed through Earth's water cycle.

▲ The rain in a hurricane comes from evaporated and condensed seawater.

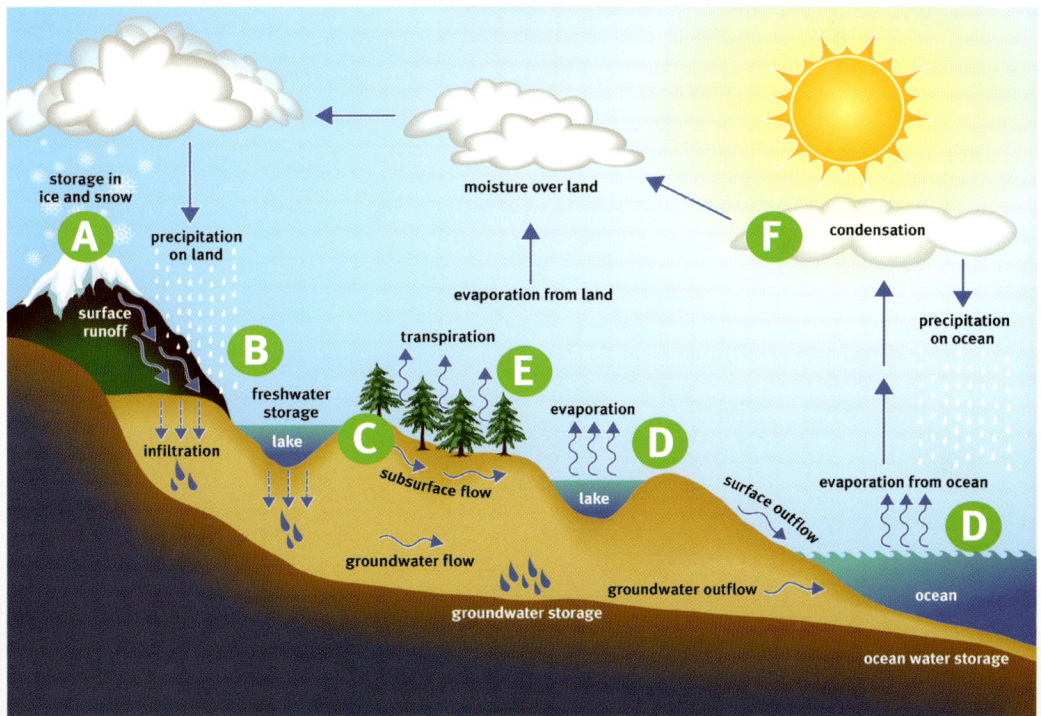

Chapter 3

HANDS-ON SCIENCE: Make Your Own Rain

You can create a tiny ecosystem and watch the water cycle in action. What creates rain?

TIME REQUIRED
30 minutes, then 15 minutes on a later day

MATERIALS NEEDED
- a clump of soil with plants
- glass or clear plastic container with a wide mouth
- hand shovel or large spoon
- plastic wrap

SAFETY CONSIDERATIONS
Be careful not to drop the glass container on a hard surface. If the glass does break, call an adult to clean up the broken glass.

Cycles in Ecosystems

PROCEDURE

1. Place the clump of soil with plants right-side up in the container. The soil should be at least 2.5 centimeters (1 inch) deep.

2. Add 125 milliliters (1/2 cup) of water to the container, or enough water to make the soil quite wet. Cover the container very tightly with clear plastic wrap. Place the container in a warm, sunny location for a few days.

3. Without moving the container, observe the changes inside it. Look for signs of condensation and precipitation in the container. Make a careful sketch of what you see.

ANALYSIS

1. Explain how the rain in the container was created.
2. Predict your results if you had not covered the container.
3. Label your drawing of the container to show the steps in the water cycle. Use the diagram on page 29 for information.

Chapter 3

THE OXYGEN–CARBON DIOXIDE CYCLE

Living communities need carbon dioxide and oxygen. These two gases are recycled in the oxygen–carbon dioxide cycle.

Producers use sunlight to convert carbon dioxide and water into oxygen and sugar. This is the process of photosynthesis. Producers use the sugar as fuel. Consumers that eat producers also use this sugar as fuel. Burning sugars to produce energy is called respiration. This process converts oxygen and sugar into energy, water, and carbon dioxide.

The waste and remains of living things fall to the ground. Decomposers eat the waste and remains. As they do, they let carbon dioxide back out into the air. Some remains were buried for millions of years. These remains change into fossil fuels. Oil, coal, and natural gas are types of fossil fuels. When people burn the fuel, the carbon is released into the air once more.

CHECKPOINT ✔

MAKE CONNECTIONS
How is the way your body makes energy similar to the way a car makes energy?

32

Cycles in Ecosystems

Summing Up

- Earth has a limited supply of the chemicals needed for life.
- Essential chemicals, such as water, nitrogen, carbon dioxide, and oxygen, are renewed through natural cycles.
- These cycles allow life to continue on Earth without using up its vital resources.

Putting It All Together

Choose one of the activities below. Respond to the prompt independently, in pairs, or in a small group. Share your responses with the class. Listen to other groups present their responses.

1. Reread Chapter 3. Which of these natural cycles depend on the actions of living things? Write a paragraph explaining your reasoning.

2. Imagine that you are a particle of water. Write a short story describing your journey through the water cycle from the ocean to the air to land and all the way back to the ocean.

3. Choose one of the natural processes described in this chapter and create a poster or diorama that shows how this process is carried out in your community.

Chapter 4
Changes in Ecosystems

How do ecosystems change over time?

▲ These bighorn rams are competing to father the next generation of lambs.

Ecosystems change every day. Sometimes it may take years to notice. Other times you may notice the change in an instant. Changes take place at every level. These changes happen over short and long periods of time.

THE STRUGGLE FOR LIFE

Ecosystems work for many reasons. Producers make food. This food feeds the community. Decomposers convert waste to help living things grow. Natural processes recycle important chemicals for reuse. The system as a whole works well together.

But each living thing struggles. Each organism fights to survive and reproduce. A flock of bighorn sheep has several rams. Each ram competes to father the next generation of lambs. The ram that survives the longest is most likely to produce the most offspring. This struggle to survive is called competition.

Chapter 4

SO MANY KINDS OF LIVING THINGS!

Over generations, populations of the same species can come to live in different environments. In time, the species grow apart and become two different species. The number of species on Earth is amazing. Scientists have discovered more than one million different types of insects alone! The species that exist today are the ones that have survived.

Each species has a different niche. For example, three different species of insect-eating birds may live in one tree. The niche of one species may be to eat insects from the top of the tree. Another species may eat insects in the middle of the tree. The other species may find food on tree trunks.

caterpillar

grasshopper

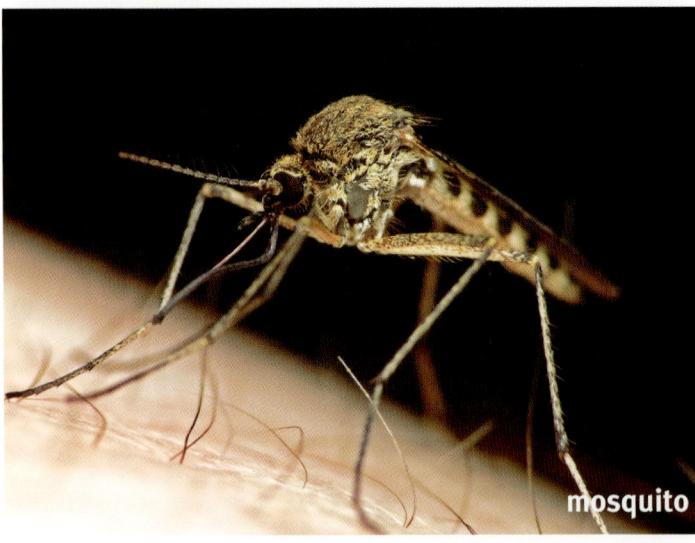
mosquito

▲ Each species fills a slightly different niche or lives in a different habitat.

CHECKPOINT ✓

READ MORE ABOUT IT
Read more about the development of new types of living things, or speciation, in your school or local library.

36

Changes in Ecosystems

Scientists have discovered more than one million different types of insects alone!

Chapter 4

SHORT-TERM CHANGES

Populations rise and fall as their ecosystem changes. Many factors can affect an ecosystem. Storms, fires, and animals can all affect an ecosystem. These factors tend to cause a short-term change. For example, locusts may invade a field. Within a few days, the insects will have eaten every green leaf they find. What effect do the locusts have on the field ecosystem?

Some niches will disappear for a time. Rabbits and other plant-eaters will move away or starve. Other niches get bigger. Snakes and other insect-eaters will eat the insects. Quick-growing weeds will thrive. Populations of some community members grow while others shrink.

In a short time, the insects run out of food and move on. Plants grow new leaves. Herbivores return. Taller plants will overgrow the weeds. Soon the ecosystem returns to its normal state.

LONG-TERM CHANGES

Sometimes whole ecosystems change slowly over many years. Your local pond may become a forest. How can this happen?

Changes in Ecosystems

The opportunities for quick-growing species of plants arise, and a new niche may be created.

▲ A swarm of locusts can change a community dramatically for a short time.

39

Chapter 4

Each year, dead plants sink to the bottom of the pond. Soil washes in from the land around the pond. The pond grows more and more shallow. Grasses and shrubs take over.

A meadow replaces the pond. Soon after, trees begin to grow there. A few more decades go by. In time you have a forest. The niches will stabilize. The period of big changes in the ecosystem will end.

Some events force an ecosystem to start over. A volcano may kill every living thing on a mountain. A glacier may leave behind bare rock. People may abandon, or leave, a strip mine.

In each case, there is no soil for plants to grow. Seeds have no place to sprout. The first living things may be lichens. These organisms start to break down the rocks. In time, bits of rock mix with lichen remains. A soil soon forms.

The new soil creates niches for other living things. Mosses and insects are carried in by wind and water. Living things from other places come to the new habitat. Slowly, the ecosystem grows. After hundreds of years, the damaged land blends in with the region around it.

▲ **This pond is well on its way to becoming a forest.**

▲ **A new ecosystem will develop on this lava flow in Hawaii.**

CHECKPOINT ✓

VISUALIZE IT
Look at the photo of the ground covered with lava. Within 100 years, the same area will be covered by low plants. Draw a picture of the lava field in 100 years.

40

Changes in Ecosystems

Summing Up

- Ecosystems are changing all the time. Individual organisms come and go.

- The organisms best suited to survive will reproduce and have offspring.

- Those not suited to survive will die and be removed from the population.

- Outside events can change ecosystems for a short time.

- Living communities bounce back quickly after storms and invasions of herbivores.

- Ecosystems can also shift gradually from one type to another.

- Over hundreds of years, new ecosystems can even develop from bare rock.

Putting It All Together

Choose one of the activities below. Respond to the prompt independently, in pairs, or in a small group. Share your responses with the class. Listen to other groups present their responses.

1. All ecosystems change over thousands or millions of years. How has your area changed since the last ice age? Research in your school or local library and present your findings in a report or poster.

2. Suppose that you take a trip to a tropical rain forest. Your friend cannot understand why there are so many different kinds of trees. Reread page 36 and think of an explanation for your friend.

3. Ecosystems contain many living things competing for resources. Malls contain many businesses competing for shoppers' dollars. Make a chart showing the niches of the various businesses in a mall.

Conclusion

Biosphere 2

▲ Biosphere 2 is an artificial environment in the Arizona desert.

Earth's environment gives us what we need to live. When people visit space, they must take many supplies. Can we create an environment that can sustain life in space? Some people are trying to answer this question.

Biosphere 2 is a huge glass building in Arizona. Biosphere 2 is named after Earth's living community, the biosphere. Biosphere 2 has five habitats: desert, rain forest, ocean, wetlands, and savanna.

Biosphere 2 has 6,500 glass panes to let in sunlight. This allows for photosynthesis to take place. Plants and algae inside the glass building make their own food. The consumers inside Biosphere 2 rely on food made by these producers.

Earth's natural cycles provide the essentials for life. Inside Biosphere 2, some of these same cycles are active. For example, plants take up water from the soil. Some of that water evaporates from leaves during transpiration and later condenses. However, the cycling within Biosphere 2 is not enough to support life. Biosphere 2 has pumps, filters, air conditioners, and other machines to help these cycles work.

Earth's ecosystems change with time. The artificial ecosystems in Biosphere 2 also change. When carbon dioxide levels rose, it affected the coral reef community. An ant invasion changed the ecology of the system for a time.

Biosphere 2 is a great achievement. But it could not support life for more than a short time. This project has helped people better understand the importance of taking care of the environment. It has taught us how special Earth is. It has also taught us how hard Earth is to copy!

▲ **Biosphere 2 contains several ecosystems that simulate Earth's natural systems.**

How to Write a Press Release

Do you have exciting news you want to share with the world? How can you get a newspaper to print an article about your news?

The answer is to send a press release to the newspaper. You might use a press release to improve attendance at an event, sell tickets, or let people know about things that are going on in the community.

Use the following steps to write a great press release.

Step 1 **Gather your information.**
Make sure you can answer the major questions about your news:
What? Who? Why? When? Where? How?

Step 2 **Write a headline.**
Make your headline short and snappy to catch the reader's attention. "Wednesday last chance for volleyball sign-up" is much better than "If you want to play volleyball this fall, you must sign up by Wednesday."

Step 3 **Write the body of your press release.**
The first paragraph answers the major questions. A second paragraph provides additional information. A last brief paragraph tells the reader how to find more information.

Step 4 **Edit your press release.**
If possible, type the article and use a word processor to check the spelling and grammar. Try to fit your information into the fewest words possible.

Step 5 **Include a photo or drawing** if you have one. A black-and-white illustration is best. Write a short caption explaining what is going on in the picture.

Step 6 **Submit the press release** to the newspaper. You can hand-deliver your press release to the school newspaper or e-mail it to the town or city paper.

Rocky the Raccoon Back in the Wild

Thursday was an exciting day for Rocky the raccoon. After three months of living in Mrs. Miller's science classroom, he ran off into the woods behind Rockvale Elementary School. Mrs. Miller's science classes adopted the orphaned cub from the Wildlife Rescue Center in February. The rescue center showed the students how to take care of Rocky until he was old enough to take care of himself.

Rocky should find plenty of food and shelter in the woods. If you see Rocky, do not approach him or offer him food. He is not a tame animal and he could bite you.

For more information, contact Mrs. Miller in Room 254 or call the Wildlife Rescue Center at 555-3456.

▲ Rocky is now roaming the woods behind Rockvale Elementary School.

Information to include in a press release:

- **What happened or will happen?**
- **Who was or will be there?**
- **Why did or will it happen?**
- **When did or will it happen?**
- **Where did or will it happen?**
- **How can the reader get more information?**

Glossary

abiotic factor (ay-by-AH-tik FAK-ter) *noun* a nonliving influence in an environment (page 11)

autotroph (AU-tuh-trofe) *noun* a living thing that makes its own food (page 16)

biosphere (BY-uh-sfeer) *noun* the portion of Earth where living things exist (page 7)

biotic factor (by-AH-tik FAK-ter) *noun* a living organism in an environment (page 11)

carnivore (KAR-nih-vor) *noun* a consumer that eats other consumers (page 18)

community (kuh-MYOO-nih-tee) *noun* all of the living things in an ecosystem (page 8)

consumer (kun-SOO-mer) *noun* a living thing that eats other living things for food (page 18)

decomposer (dee-kum-POH-zer) *noun* a living thing that eats the wastes and dead bodies of once living things (page 18)

46

ecology	(ih-KAH-luh-jee) *noun* the study of the relationships among living things and their interaction with the nonliving environment (page 8)
ecosystem	(EE-koh-sis-tum) *noun* all the interacting living and nonliving things in one area (page 8)
food web	(FOOD WEB) *noun* a diagram of the interconnecting food chains between producers and consumers in an ecosystem (page 21)
habitat	(HA-bih-tat) *noun* the place where a living thing makes its home (page 10)
herbivore	(ER-bih-vor) *noun* a consumer that eats only plants (page 18)
niche	(NICH) *noun* the survival strategy of a species of living thing (page 12)
omnivore	(AHM-nih-vor) *noun* a consumer that eats both producers and consumers (page 18)
population	(pah-pyuh-LAY-shun) *noun* all the members of one type of living thing in the same area (page 10)
producer	(pruh-DOO-ser) *noun* an autotroph that produces food for those that cannot (page 16)

Index

abiotic factors, 11, 13
atmosphere, 4, 7–8, 27–29
autotroph, 16
biomass, 22, 24–25
biosphere, 7–8, 13, 43
Biosphere 2, 43
biotic factors, 11
carnivore, 18, 24–25
community, 8, 10, 21–22, 35, 38, 40, 43
consumer, 18, 21–22, 25, 27, 32, 43
decomposer, 18, 25, 28, 32, 35
ecology, 8, 13, 43
ecosystem, 8, 11, 13, 16, 21–22, 35, 40–41, 43
energy, 16, 18, 21–22, 25, 32, 43
environment, 4, 8, 11, 13, 27, 36, 43
food chain, 21, 24–25

food web, 21–22, 25, 28
fossil fuels, 32
Galdikas, Biruté, 20
habitat, 10, 12–13, 20, 36, 40, 43
herbivore, 18, 24–25, 38, 41
International Space Station, 4, 7
niche, 12–13, 36, 38, 40
nitrogen cycle, 28
omnivore, 18, 25
oxygen–carbon dioxide cycle, 32
photosynthesis, 16, 32, 43
population, 10, 13, 35–36, 38, 41
producer, 16, 18, 21–22, 24–25, 27, 32, 35, 43
respiration, 32
transpiration, 29, 43
water cycle, 29

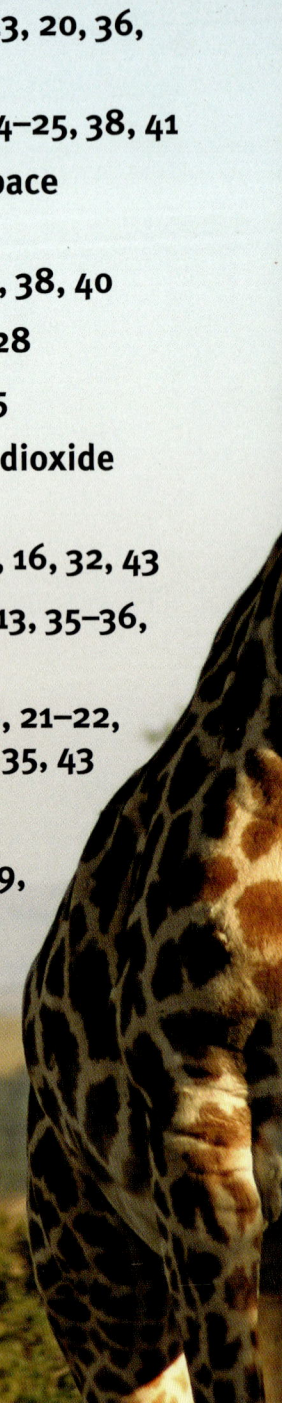